34134 00057423 3
Leabharlainn nan Eilea
KU-071-226
WITHDRAWN

LIFE SCIENCE IN DEPTH

VARIATION AND CLASSIFICATION

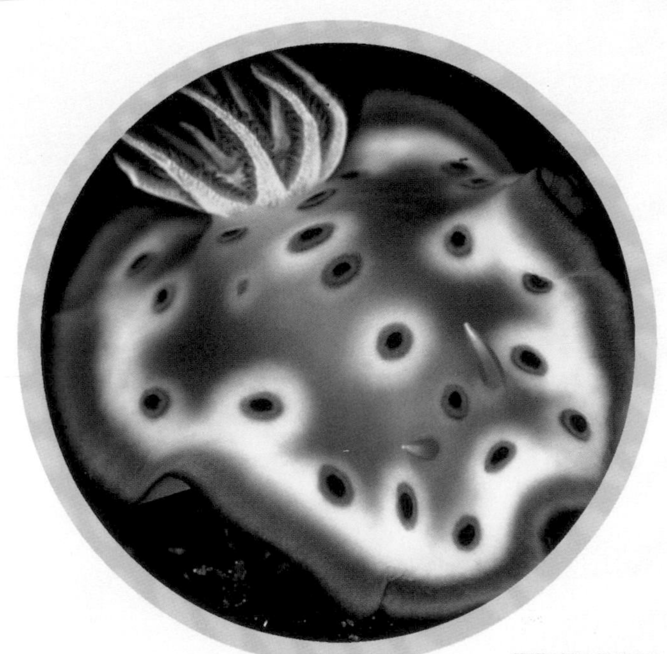

Ann Fullick

www.heinemann.co.uk/library
Visit our website to find out more information about Heinemann Library books.

To order:

 Phone 44 (0) 1865 888066
 Send a fax to 44 (0) 1865 314091
 Visit the Heinemann bookshop at www.heinemann.co.uk/library to browse our catalogue and order online.

First published in Great Britain by Heinemann Library, Halley Court, Jordan Hill, Oxford OX2 8EJ, part of Harcourt Education.

Editorial: Sarah Shannon and Dave Harris
Design: Richard Parker and Q2A Solutions
Illustrations: Q2A Solutions
Picture Research: Natalie Gray
Production: Chloe Bloom

Originated by Modern Age Repro
Printed and bound in China by South China Printing Company

10 digit ISBN: 0 431 10900 1
13 digit ISBN: 978 0 431 109008

10 09 08 07 06
10 9 8 7 6 5 4 3 2 1

British Library Cataloguing in Publication Data
Fullick, Ann, 1956-
Variation and classification.
- (Life science in depth)
576.5'4
A full catalogue record for this book is available from the British Library.

Acknowledgements
The publishers would like to thank the following for permission to reproduce photographs: Alamy pp. **6** (Andrew Fox), **34** (Foybles), **43** (Gary Meszaros/Bruce Coleman), **24** (Juniors Bildarchiv); Botany Library of Harvard University p. **31**; Corbis pp. **7** (Bettmann), **28** (Dennis Degnan), **51** (E.O. Hoppe), **27** (Hulton-Deutsch), **52** (Louie Psihoyos), **16** (Melissa Moseley/Sony Pictures/Bureau L.A. Collections), **41** (Paul A Souders), **22** (Robert Pickett), **45** (Tom Brakefield), **22**; Digital Vision pp. **19**, **48**, **55**; FLPA p. **4** (Albert Visage); Getty Images pp. **56** (John Giustina), **25** (PhotoDisc); Harcourt Education Ltd pp. **21a**, **21b**, **21c**; Mary Evans p. **32**; Natural History Museum p. **46**; Oxford Scientific pp. **1**, **30**; Science Photo Library pp. **9a**, **9b** (CNRI), **8** (Dr Gopal Murti), **15** (James King-Holmes), **11** (Philippe Plailly/Eurelios), **38** (Volker Steger/Christian Bardele); Trustees, Royal Botanic Gardens, Kew p. **59**.

Cover photograph of tropical fish, reproduced with permission of Digital Vision.

Our thanks to Emma Leatherbarrow for her assistance in the preparation of this book.

The paper used to print this book comes from sustainable resources.

Contents

Words printed in the text in bold, **like this**, are explained in the Glossary.

The variety of life

Look around you – whether you live in an apartment in the city or in a cottage out in the countryside, you are surrounded by more living organisms than you can possibly imagine. When you look further, to the distant corners of planet Earth, the variety becomes mind-boggling.

THE PLANTS

There are about 250,000 known **species** of flowering plants in the world – and lots more yet to be discovered. They come in all shapes and sizes, ranging from absolutely huge to microscopic. The flowers of *Rafflesia arnoldii*, found in the rainforests of Indonesia, measure about 4 metres (13 feet) around the edge of the petals, and stink of rotting meat to attract the flies needed to **pollinate** it. At the other end of the scale, the tiniest flowering plant, *Wolffia globosa*, (a tropical duckweed) is only about 0.6 millimetres (0.024 inches) long

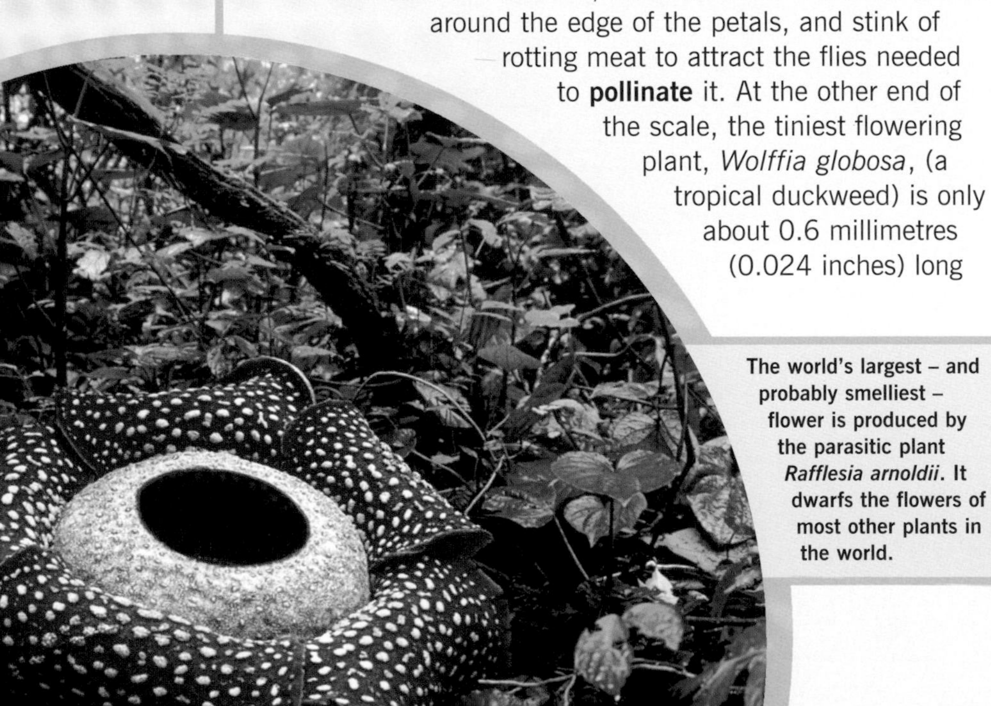

The world's largest – and probably smelliest – flower is produced by the parasitic plant *Rafflesia arnoldii*. It dwarfs the flowers of most other plants in the world.

and 0.3 millimetres (0.012 inches) wide. It weighs about the same as two ordinary grains of table salt – seven trillion times lighter than one of the giant sequoia trees found in the United States! Of course, as well as the flowering plants, there are many thousands of organisms, like ferns and mosses, that do not produce flowers but are still plants.

THE ANIMALS

Animals are another source of endless variety. From microscopic animals that live in the sea and oceans, to the elephants of Africa and Asia, from the seals and penguins that can survive in some of the coldest places on Earth, to the lizards and camels that thrive in the hottest deserts. Although animals cannot grow as large as plants because they have to be able to move about, the number of different lifestyles, body shapes, and colours runs into hundreds of thousands.

CLOSE TO HOME

Plants and animals are not the only types of living organisms. Your own surroundings will be full of different types of living things – often incredibly small. For example, the mattress of your bed will be home to some very small animals – between 100,000 to 10 million house dust mites will be living there! In there with them, feeding on your dead skin are **fungi** such as yeast as well as hundreds, maybe thousands, of different types of **bacteria**.

Did you know..?

The huge variety of living organisms that share our homes can affect our health. An allergy to house dust mites or their droppings can trigger **asthma**. Although house dust mites are tiny, there are lots of them. About 10 percent of the weight of a two-year-old pillow is made up of dead house dust mites and their droppings!

SPOT THE DIFFERENCE!

It is easy to identify a tree, a person, or a herd of cows. However, it is not always so easy to see the differences between the cows in a herd, the trees in a wood, or even the people in a crowd. Yet although these differences may not be so easy to spot, they are there, and they are important. These small variations make every individual organism unique. These differences come from the genetic information that is passed on from parent to offspring, and from the environment in which an organism grows up. As **naturalists** have built up their collections of living things, the differences both between species and within species have become clearer.

If a group of people who all own the same breed of dog get together, everyone will know their own dog. In the same way, babies tend to look quite similar – unless you are a parent, when you can spot your own very easily!

SCIENCE PIONEERS Charles Wilkes and the Exploring Expedition of 1838–1842

In the 17th and 18th centuries, people started to travel the world, and they discovered more and more of the great variety of life. It took great courage to set out for unknown lands in the sailing ships of the time to draw maps and collect specimens. One of these explorers was the American Charles Wilkes who set off in 1838 with six ships and 346 men, including plenty of scientists and naturalists. Four years later, the expedition had lost two ships, travelled 140,000 kilometres (87,000 miles), surveyed 280 islands in the Pacific Ocean, and discovered a huge number of plants and animals. Wilkes brought back specimens of about 10,000 species of plants and 4,000 animals. Almost half of the animal specimens alone were new species.

When Charles Wilkes set off in 1838, he can have had no idea that he was going to discover such a tremendous variety of life.

SORTING THINGS OUT

As explorers and collectors have brought back their specimens over the last few hundred years, the number of known types of living things has grown and grown. In fact it continues to grow – we are still discovering new species all around the world almost every day. For many years, scientists have tried to make sense of this great variety of life. Some have organized the huge numbers of different organisms, putting them into groups that everyone can refer to. Others have worked to understand how this great variety has come about, and what causes the differences both between different species, and between the members of the same species. By looking at all their work, we too can begin to understand the riches of the natural world.

Variety in the genes

In every family there are some features that clearly show a "family likeness". It might be a particular shaped nose, or the ability to wiggle the ears. Sometimes, when old family portraits exist, family likenesses can be seen between people living today and their ancestors from centuries ago. Characteristics are passed on (or inherited) from one generation to another. Yet it is only in the last century that we have finally understood how **inheritance** works, and where variety comes from.

THE INFORMATION IN THE CELLS

Your body is made up of billions of cells, and most of those cells contain the instructions needed to make a new you. Each cell contains a **nucleus**, which stores the information you inherited from your parents – the blueprint for making you.

People produce babies by **sexual reproduction**. Special sex cells called **gametes** join together to form a new individual. The sex cell from the mother is called an **ovum** (egg) and the sex cell from the father is the **sperm**. These sex cells

Human cells vary greatly in size but what most of them have in common is that they contain a nucleus (the red area in this picture). The nucleus contains genetic information.

carry information about a person's characteristics. You will inherit some characteristics from your mother and some from your father, but you won't be exactly like either of them.

The information is stored in the nucleus of the **chromosomes**. Chromosomes are made up of an amazing chemical called **DNA** (which stands for **deoxyribonucleic acid**). DNA carries the instructions to make all the **proteins** in your cells. Many of these proteins are **enzymes**, which in turn control the production of all the chemicals that make up your body and affect what you look like and who you are.

WHAT'S YOUR NUMBER?

Each different organism has a set number of chromosomes in each normal body cell. Humans have 46 chromosomes. Carrots have only 18, while alligators have 32 and horses have 64. Chromosomes come in pairs and you inherit one half of each pair from your mother and the other half from your father. So, people have 23 pairs, carrots 9 pairs, alligators 16 pairs, and horses 32 pairs of chromosomes.

When the chromosomes are dividing, they become shorter and thicker. Scientists can photograph them and arrange the images to make a special picture known as a karyotype that shows the size and shape of all the different chromosome pairs in the cell. This is a female karyotype on the left, and a male karyotype on the right.

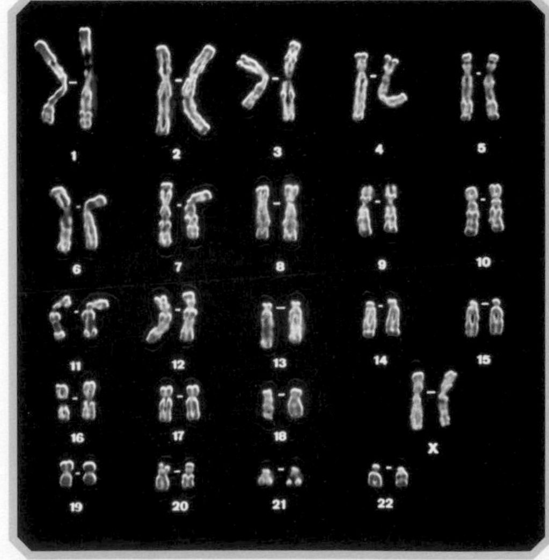

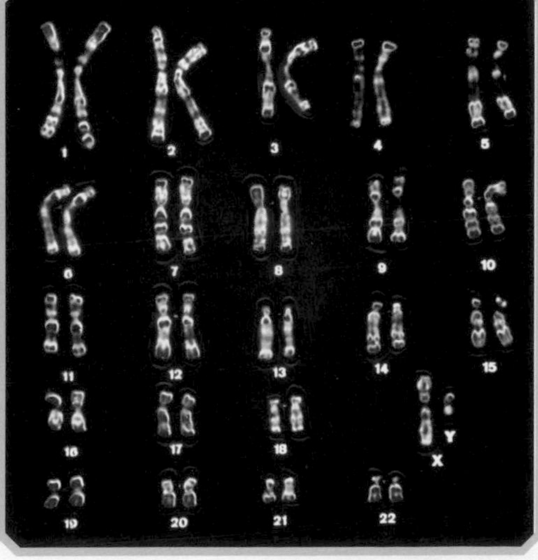

A NEW CELL, A NEW PERSON

When the gametes are made, the chromosome pairs split up. So each human egg has 23 chromosomes, and so does every sperm. When the egg is fertilized by a sperm and they join together, the new cell, which will grow and divide to form a new person, contains the full 46 chromosomes it needs.

GIRL OR BOY?

In 22 of the chromosome pairs, both chromosomes are the same size and shape in both men and women. They are called the **autosomes**. They control almost everything about the way you look and the way your body works. The remaining pair of chromosomes are the **sex chromosomes** – they decide whether you are male or female. There are two types of sex chromosomes, which are different in size and shape. One of the pair, called the **X chromosome**, is the same size as the autosomes. The other one, called the **Y chromosome**, is shorter. Everyone inherits one X chromosome from their mother. If this joins with a sperm carrying another X chromosome, you will be a girl (XX). If it is fertilized by a sperm carrying the Y chromosome, you will be a boy (XY).

X chromosomes carry information about female characteristics, but they also carry information about other things like the way your blood clots and the formation of your teeth and hair. Y chromosomes carry information about male characteristics and just a few other bits of information, including some about hairy ears!

DNA – THE MOLECULE OF LIFE!

All the information needed to make a new you is carried in your **genes**. Each gene is a small section of DNA, a long **molecule** made up of two strands that are twisted together to make a spiral called a **double helix**. The DNA molecule is made up of smaller molecules joined together. These include four molecules called **bases**, which appear time after time in different orders but always paired up in the same way. Genes are made up of repeating patterns of these bases. Guanine always pairs up with cytosine, while adenine pairs up with thymine.

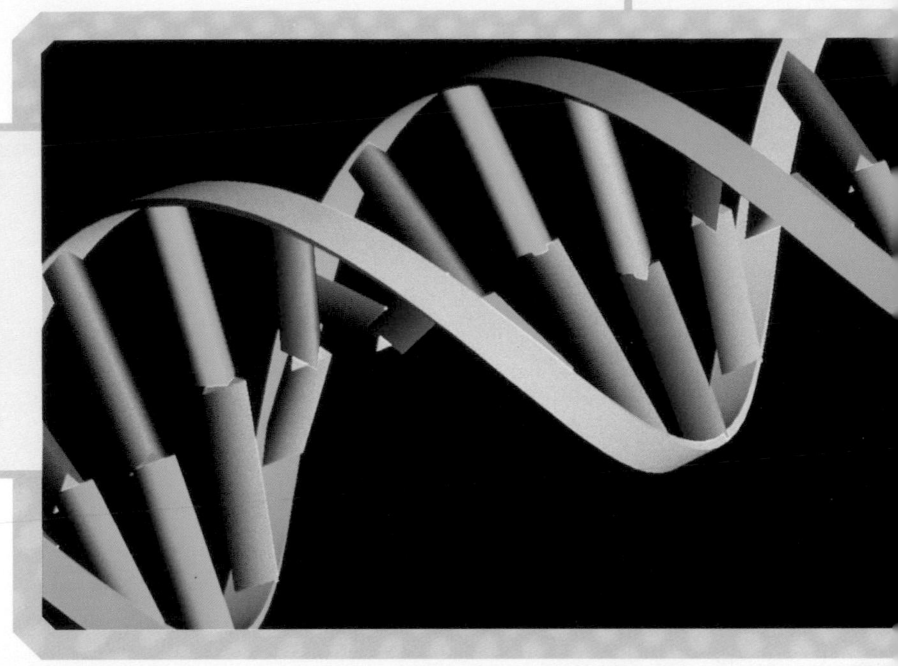

This is a computer generated image of the structure of a DNA molecule. A tiny change in the arrangement of the bases in your DNA (shown here as the different-coloured bars) would have meant a completely different you!

SCIENCE PIONEERS
Watson and Crick vs. Wilkins and Franklin

In the 1950s, there was a race between two teams of scientists to be the first to unravel the secrets of the DNA molecule.

Maurice Wilkins and Rosalind Franklin in London took special X-ray photographs of DNA and looked for patterns in the X-rays that would show them the structure of the molecule. At the same time, James Watson (a young American), and Francis Crick (from the UK) were working on the same problem at Cambridge. They took all the information they could find on DNA – including the X-ray pictures from London – and tried to build a model of the molecule that would explain everything they knew. Once they realized that the bases always paired up in the same way, they had cracked the DNA code, and built the now famous DNA double helix for the first time!

THE GENE LOTTERY

Each of your chromosomes carries thousands of genes. They are arranged so that both chromosomes in each pair carry genes controlling the same things in the same positions. So your genes also come in pairs, half from each parent.

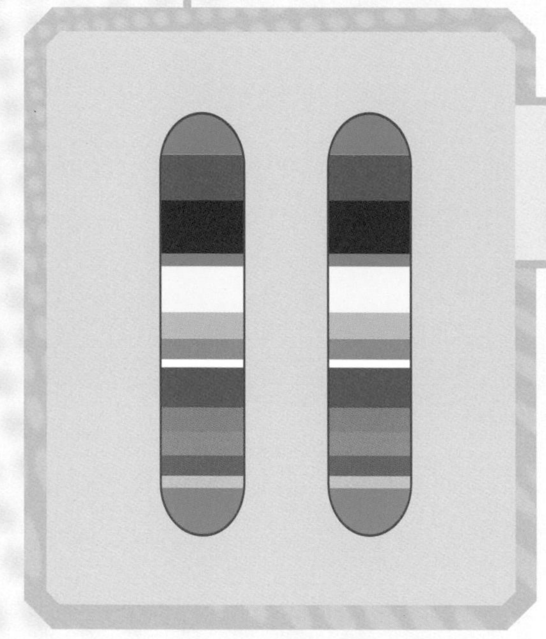

The pairs of genes on the chromosomes interact to control most things about the body, and a surprising amount about the mind too. Genes are shown as coloured bands on this chromosome diagram.

ALLELES

Each pair of genes controls different things about you. Half of each pair of genes came from your mother (in the egg cell), and half from your father (in the sperm). However, the genes in each pair can come in different forms. These different versions of the same gene are called **alleles**. For example, there is a gene that decides whether your thumb is straight or curved. One possible allele gives instructions for a straight thumb while the other is for a curved thumb. We can use these genes to help us understand how inheritance works.

The genetic information in each sex cell is a unique mix, so when the gametes from two different people join together at fertilization, the new cell formed is also completely unique. This is where much of the variety of life comes from. However many children there are in a family, they will never be exactly the same (unless they are identical twins, because identical twins are formed from just one original fertilized egg, so their DNA is the same).

We can use a simple model to help us understand how we inherit different combinations of alleles from our parents. Imagine a bag containing a mixture of red and blue marbles. If you put your hand in and, without looking, picked out two marbles at a time, what combinations might you get? You could end up with three possible pairs – two blue marbles, two red marbles, or one of each colour. This is what happens when you inherit genes from your parents. For example, if both of your parents have two alleles for dimples (like the red marbles) you will definitely inherit two dimple alleles and you will have dimples. If both of your parents have two alleles for no dimples, you will inherit alleles for no dimples and you will be dimple free. But if your parents both have one allele for dimples and one for no dimples, you could end up with two dimple alleles, two no dimple alleles – or one of each!

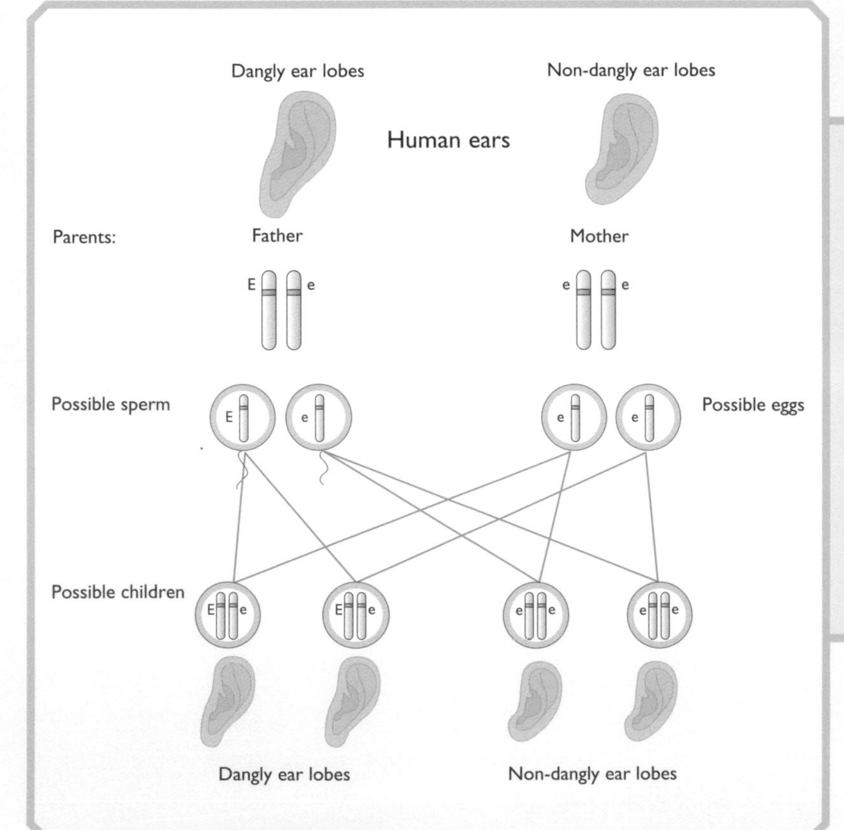

The allele for dangly ear lobes is dominant, which is shown by the capital letter E in the father's chromosomes. However, the father has one recessive allele for non-dangly ear lobes. If this combines with one from the mother, the child will have non-dangly ear lobes.

UNDERSTANDING GENETICS

Genetics is not just about people. The same ideas apply to living things from elephants to jellyfish, from giant redwood trees to moss. The numbers of chromosomes may vary but whenever sexual reproduction takes place, genetics is at work. Yet for hundreds of years people had no idea how information was passed from one generation to another. We owe our understanding of genetics to an Austrian monk called Gregor Mendel and his experiments with peas.

SCIENCE PIONEERS Gregor Mendel

Gregor Mendel was born in 1822 in Heinzendorf, Austria. He was incredibly clever – but very poor. In those days, the only way for a poor person to get an education was to join the church, so Mendel became a monk. He worked in the monastery gardens and became fascinated by the way things like colour and skin wrinkliness were inherited by the growing peas there. He decided to carry out some breeding experiments, using pure strains of round peas, wrinkled peas, green peas, and yellow peas for his work. Mendel cross-bred the peas and made a careful note of the type of plants produced. He counted the different offspring carefully and found that characteristics were inherited in clear and predictable patterns. Mendel explained his results by suggesting there were separate units of inherited material. He realized some characteristics were **dominant** over others and that they never mixed together. This might not sound very exciting to us, but to Mendel 150 years ago it was thrilling. The Abbot of the monastery was also interested in Mendel's ideas and built him a large greenhouse for carrying out his experiments.

Mendel kept records of everything he did, and analysed his results – something almost unheard of in those days. Finally in 1866, Mendel published his findings. He explained some of the basic laws of genetics in a way we still use today. Sadly Mendel's genius fell on deaf ears. He was ahead of his time – no-one yet knew of the existence of genes or chromosomes, so people simply did not understand his theories. He died eighteen years later with his ideas still ignored, but he was sure that he was right.

Sixteen years after his death, Gregor Mendel's work was finally recognized. By 1900, people had seen chromosomes through a microscope. Three scientists, Hugo de Vries, Erich von Seysenegg, and Carl Correns discovered Mendel's papers and duplicated his experiments. When they published their results, they gave Mendel the credit for what they observed. From then on, ideas about genetics developed rapidly. It was suggested that Mendel's units of inheritance might be carried on the chromosomes seen beneath the microscope, and the science of genetics as we know it today was born.

This is a portrait of Gregor Mendel. His ideas on genetics were years ahead of his time.

MUTATION!

Sexual reproduction is the main powerhouse behind the variety of life on Earth – but variety also comes about as the result of mutation, when changes take place in the DNA. Most people think of **mutation** as dramatic, causing huge, scary changes. In fact, it happens all the time. The vast variety of organisms today is the result of billions of mutations, which have taken place since the earliest life on Earth.

A mutation is a change in a gene – the appearance of a new allele as the result of tiny changes made in the long strands of DNA. Mutations often happen naturally when mistakes are made in copying the DNA for new cells. When mutations take place in the sex cells, they can be passed on to the next generation.

Some things increase the chances of mutations happening. If your cells get lots of radiation from **radioactive** substances, **ultraviolet** light from the Sun or X-rays, mutations are more likely to happen. Some chemicals can also cause mutations, and there are lots of these in cigarette smoke.

Many mutations do not have any effect at all because they take place in parts of the DNA that do not control what an organism looks like. Some mutations are very harmful though. In most cases, an **embryo** that inherits a harmful mutated gene will die at an early stage of pregnancy, although some keep developing and are born with a **genetic disease**.

Mutation of the DNA in fiction is fun – but often not very realistic!

Just occasionally, a mutation can produce something that helps an organism to survive more easily. This increases its chances of breeding and passing on the good mutation, adding to the variety of life. If the mutation is really useful, it will become more and more common.

RECENT DEVELOPMENT
The Human Genome Project

In the late 20th century, scientists set up the Human Genome Project to identify all of the genes in the human chromosomes – and then to discover the 3 billion base pairs which make up the human DNA! The project was a huge international effort. Scientists in eighteen different countries were working on different bits of the **genome** at the same time. They announced the human genome in the year 2000. The Project cost around 2.7 thousand million US dollars, and has shown that in spite of all the variety in people, every individual has at least 99.99 percent of their DNA in common!

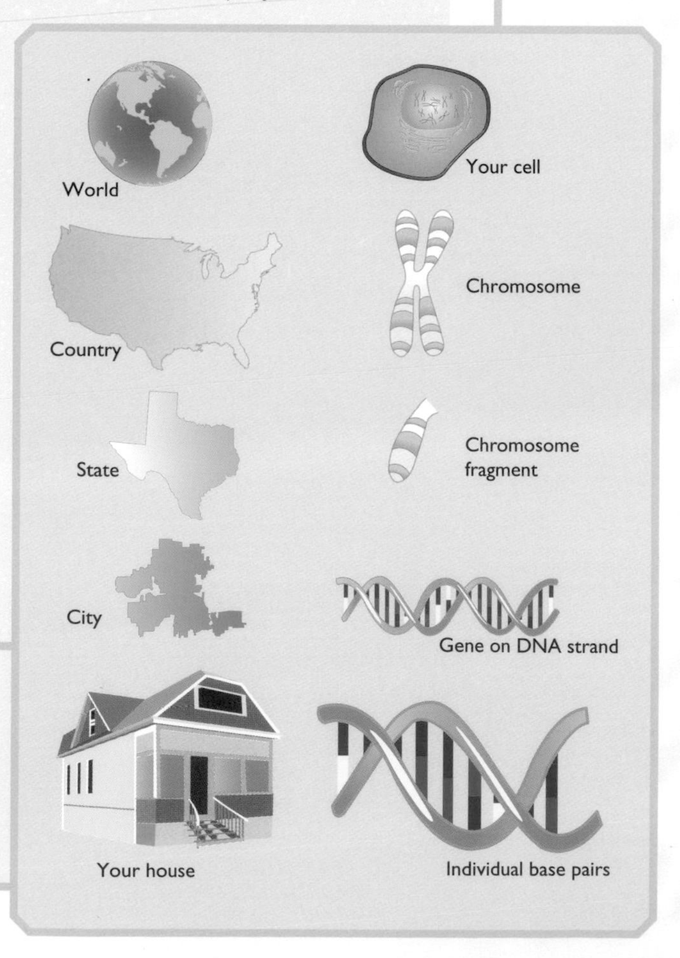

Working out the bases of the human genome is a bit like taking a map of the world and pinpointing your personal address. Eventually it will make it possible for you to know your own genetic makeup.

Variety from variety!

There are millions of different species of living organisms in the world today. Scientists explain the variety of life as the result of a process called **natural selection**. The natural world is a harsh place and living things are always in competition with each other for food, water, and space to live. When a chance mutation (see page 17) causes a helpful change, the organism gains an advantage in the competition against other species and against other members of its own species. So individuals with the new characteristic are more likely to survive and breed.

SURVIVAL OF THE FITTEST

Charles Darwin was the first person to explain natural selection by the "survival of the fittest" but what does this mean? Reproduction is a very wasteful process. There are always more offspring than the environment can support. But not all the offspring survive – it is the offspring with the genes best suited to life in their **habitat** that will be most likely to stay alive and breed successfully. This is natural selection.

For example, rabbits with the best eyesight, the sharpest ears, and the longest legs will be the ones that are most likely to escape being eaten by a fox. They will hear or see the fox coming, and they can run away fast! Those rabbits will be the ones most likely to survive and breed, and they will pass those useful genes on to their babies. The slower, less alert rabbits will get eaten – so their genes will be digested along with the rest of them! Over time, this process leads to animals being better suited to their habitat.

SCIENCE PIONEERS Charles Darwin

In 1831, a 22-year-old Charles Darwin started five years adventuring on the *HMS Beagle*. By the end of the voyage, Darwin had collected so many plant and animal specimens, and made so many drawings and notes, that he had enough work identifying them all to last him for years. He was fascinated by the way that animals on each of the Galapagos islands, which were obviously closely related, like the giant tortoises, all differed slightly because they were adapted to take advantage of the slightly different habitats on each island. When he returned home, it took over 20 years of studying specimens and thinking about his ideas before Darwin finally published his now-famous book, *On the Origin of Species by Means of Natural Selection*.

The amazing marine iguanas from the Galapagos Islands are very like the inland iguanas, except that they are specially adapted to cope with spending part of each day in the sea, eating seaweed off submerged rocks. Their dark colour allows them to absorb lots of heat from the Sun to warm them up before and after being in the water, and camouflages them on the black rocks.

THE SAME, BUT DIFFERENT...

At least part of the variety around us seems to be due to differences in the environment where organisms live. Variety that is the result of genetic differences can be passed on from parents to their offspring. Variety that is the result of lifestyle or habitat differences cannot be passed on – but it can have a big effect.

ENVIRONMENTAL FACTORS

Imagine a patch of nettles growing in the bright sunshine. They have everything a plant could need to grow – plenty of water, warmth, and most of all, plenty of light. Now imagine a similar patch of nettles growing in deep shade under some bushes. Even if the two clumps of plants are genetically identical, they will not be the same. You might expect the plants in the light to be the biggest, but in fact the nettles growing in the shade will probably be taller and probably have bigger leaves as well.

This is because the plants in the shade are fighting hard to get more light. Growing upwards increases their chances of breaking through the covering plants and reaching the light. Plants need to catch as much light as possible to **photosynthesize** successfully. Large leaves have a big surface area to make the most of any light available. If the nettles in the sunshine were moved into the shade, they would soon look like the plants that had been shaded all along. So light levels will affect the appearance of a plant regardless of its genetic makeup.

It is not just light that affects plants either. If they are overcrowded, they will grow tall and spindly as each one tries to outgrow the others and get the most light. In exposed conditions, plants become small and stunted as they try to minimize the damage done by the wind and cold.

KEY EXPERIMENTS
the mineral requirements of plants

Plants make their own sugars from water and carbon dioxide using light in a process called photosynthesis. However, they also need minerals from the soil to turn the sugars into proteins and other chemicals that they need. There is a classic experiment that shows clearly the effect of different minerals on plant growth and appearance.

A number of similar young plants are grown in water that contains everything they need except for one mineral such as nitrate, phosphate, potassium or magnesium. The plants are kept warm and given plenty of light, and their growth and appearance is checked regularly. Any changes in how they look are down to the lack of a mineral, not differences in the DNA.

When you see how minerals – or the lack of them – affects plant growth it is easy to understand why farmers and gardeners keep their soil well supplied with mineral rich fertilizers.

ANIMALS CHANGE TOO

Variety in animals, just like that in plants, can be closely linked to the conditions around them. Some of the biggest changes in an animal's surroundings come with the changing of the seasons. A habitat full of greens and browns in the spring and summer can be completely white with frost and snow through the winter months. This means that animals that are perfectly camouflaged at one stage of the year can be very obvious to **predators** at another. To deal with this problem, some animals hibernate (go into a deep sleep) through the winter. Others change the colour of their fur or feathers depending on the time of year.

The trigger that causes animals to change their coats is usually either a change in the day length or in the average temperature, or a combination of both. The fur or feathers are formed with or without colour as a result of alterations in the animal's body chemistry. For example, the Arctic fox changes from grey-brown in the summer to white in the winter.

The effect of the environment on the colour of this Arctic fox in summer and winter is plain to see.

In a laboratory, scientists can see what happens to animals when they keep them in artificial conditions. If the conditions stay the same (for example, no change in day length or temperature) then the normal, seasonal changes in coat colour simply do not occur. So, although these animals have the genetic information that makes it possible for them to change, it does not happen unless the environmental triggers are there as well.

SINGING THE RIGHT SONG

If you stand outside and listen, you will almost always hear the sound of birds, from the cheeping of sparrows and the cooing of city pigeons to the songs of many different birds in the countryside. Songbirds rely heavily on their songs to pass on all sorts of information and most have their own, very typical song. Experiments have shown that if baby birds are brought up without hearing the song of their parents, then they do not learn to sing properly. They hatch out of the egg with the ability to recognize their own type of song and learn it – but they need to hear the song around them to actually do it. The environment affects the way they develop.

Just like with plants, a whole range of environmental factors, including light, temperature, food, and sound, can have a big impact on an animal, regardless of its genetics. No one is quite sure about which of these is the most important factor though – scientists have argued about it for years!

Nature versus nurture

The way an organism develops depends on its genetic information, but is there more to it than that? For many years, people have been trying to sort out the importance of nature (the genes you inherit) versus nurture (the conditions in which you grow). In the 21st century the debate continues!

NATURE – IT'S ALL DOWN TO THE GENES

The genes we inherit decide a great deal about us. An apple seed will never grow into an oak tree, regardless of the weather conditions or the soil in which it is planted. Elephants always give birth to baby elephants and cod always spawn baby cod. The basic characteristics of a species are decided by the genes it inherits from its parents. As our knowledge of genes and genetics increases, the evidence for the importance of our genes gets stronger and stronger. For example, scientists now know that not only are genes responsible for genetic diseases, but they also have a part to play in deciding whether a person will develop problems such as heart disease and cancer later on in life. However, genes are not the whole story.

No matter how much these dogs eat, or how good the conditions in which they are kept, they will never grow as tall as an elephant – it just is not in their genes!

NURTURE – THE ROLE OF THE ENVIRONMENT

While the genes play a vital part in deciding just what an organism will look like, the conditions in which it grows and develops are also very important. If genetically identical plants are grown under different conditions, the effect of the environment shows up clearly. What is more, genetically identical – or at least very similar – animals can now be produced in laboratories. This allows scientists to learn more about the effect of the environment on the way animals develop. Furthermore, just as some research has shown that your genes affect your future health, other research shows that the environment you live in – things like the food you eat, the amount of exercise you take, whether you smoke, and how much you drink – also has a big impact on the diseases you are likely to develop.

SCIENCE PIONEERS Sir Richard Doll and the effect of smoking

Over 50 years ago, a scientist called Richard Doll was trying to explain the huge increase in lung cancer in the UK. He thought it could be due to something in the environment, and suspected that it might be the tar used to make roads. In fact, his research showed clearly that cigarette smoke was causing the damage.

Sir Richard Doll demonstrated a clear link between environment and health, which has since saved thousands of lives.

THE HUMAN QUESTION

Plants and animals are very useful as models for how genes work and the influence of the environment. It is relatively easy to work with thousands of genetically identical plants, or with animals like fruit flies, which reproduce in a matter of days. You can look at how they mutate, and grow them under different conditions to find out what affects them. But there is a limit to what these models can tell us about people. Unfortunately, it would be completely impossible to do the same type of experiments on human beings!

People cannot be used as laboratory specimens for lots of reasons. We only have a small number of children – usually one at a time – and it takes nine months for each baby to grow and develop. More importantly, people choose their own partners. A scientist cannot ask two people to have a baby just to see what the offspring of a particular cross might look like! Also it would be completely **unethical** to bring children up in bad conditions just to see how they grow and develop. So how do we find out about the sort of things that affect human beings, and the differing roles of nature and nurture?

COLLECTING THE FACTS

One thing scientists can do is look at what happens when genetic information goes wrong. Studying genetic diseases, such as **cystic fibrosis** and **haemophilia**, has helped scientists find out a great deal about the way normal genes work and what they control.

Another way of learning more is to collect information about very large numbers of people and see what appears to affect their growth and health. This is the sort of work carried out by Sir Richard Doll in his study on the effects of cigarette smoke (see page 25). For example, the amount and type of food you eat affects how you grow. Children given a poor diet will not grow as tall as they would if they had a good, balanced diet.

About 60 years ago, during the Second World War, food in the UK was rationed and there was not as much to eat as there is today. Information collected over the years shows that since the war ended, children have been getting taller in the UK as food has become more plentiful. In England, boys are on average 0.5 centimetres (0.2 inches) taller now than they were in the 1940s. This shows that having different diets is one of the factors that can cause variations among whole populations of people.

Rationing during the 1940s meant the protein and calcium children need to grow was in short supply. As a result, they did not grow as tall as they might have done with plenty of food available.

TWIN STUDIES HOLD THE KEY

Identical twins are the only human beings with exactly the same genes. They are formed when a fertilized egg splits completely in two. Identical twins give us a natural opportunity to find out the effect on a child of the environment it grows up in. When identical twins are born, you might expect that they would weight exactly the same, because they are genetically identical. But it is actually very unusual for identical twins to be the same weight when they are born. Although they have both been in the same **uterus**, sharing the same **placenta**, they have been in different positions and sometimes one has had a better blood supply than the other.

It would be unethical to take identical twins away from their parents and bring them up in different ways just to see what effect it has. However, over the years there have been a number of identical twins who were adopted by different families. Some of them did not even know that they had a twin until later on in life!

Scientists have researched this, tracing twins who have lived very different lives and looking at the similarities and differences between them. The similarities in appearance between twins when they meet up – choice of clothes, hairstyles, and so on –

Identical twins give scientists a chance to try and find out which of our characteristics depend only on our genes, and which are affected by the world in which we grow up.

are often quite amazing. Some of the features of the twins, such as their height, weight, and IQ, can be measured scientifically. By doing this, scientists have shown that in humans, just like other organisms, some of the differences between us are due to genetics, some are due to environment, and some are due to a bit of both. But the more evidence that is studied, the bigger the part the genes seem to play. Nature seems to be winning over nurture.

KEY EXPERIMENT tracking twins

To see the effect of being brought up in different environments, scientists have compared the average height, weight, and IQ from a group of separated identical twins with those of a group of identical twins who were brought up together, and with those of a group of brothers and sisters who were not twins. The identical twins brought up together were most alike, but identical twins brought up in completely different homes were more alike than normal brothers and sisters. The results showed clearly that the environment you are brought up in certainly does have an effect on how you develop, but so do your genes.

The average difference measured between twins or siblings	Identical twins brought up together	Identical twins brought up apart	Non-twin siblings
Height (cm) average difference in height	1.7	1.8	4.5
Weight (kg) average difference in weight	1.9	4.5	4.7
IQ points (a measure of intelligence) average points difference in IQ	5.9	8.2	9.8

Making sense of the world

The variety within the living world is enormous, yet for thousands of years people were unaware of it. Each group of people only knew about their own small area, and while there is plenty of variety even in a few metres of grassland or woodland, it does not compare to the variety of organisms around the world. Once travelling between countries and continents became possible, people became more aware of the richness of the living world. Naturalists and biologists have since travelled the globe to find as many different types of living things as possible.

The amazing colours of the nudibranch molluscs are just one example of the variety of life discovered under the sea.

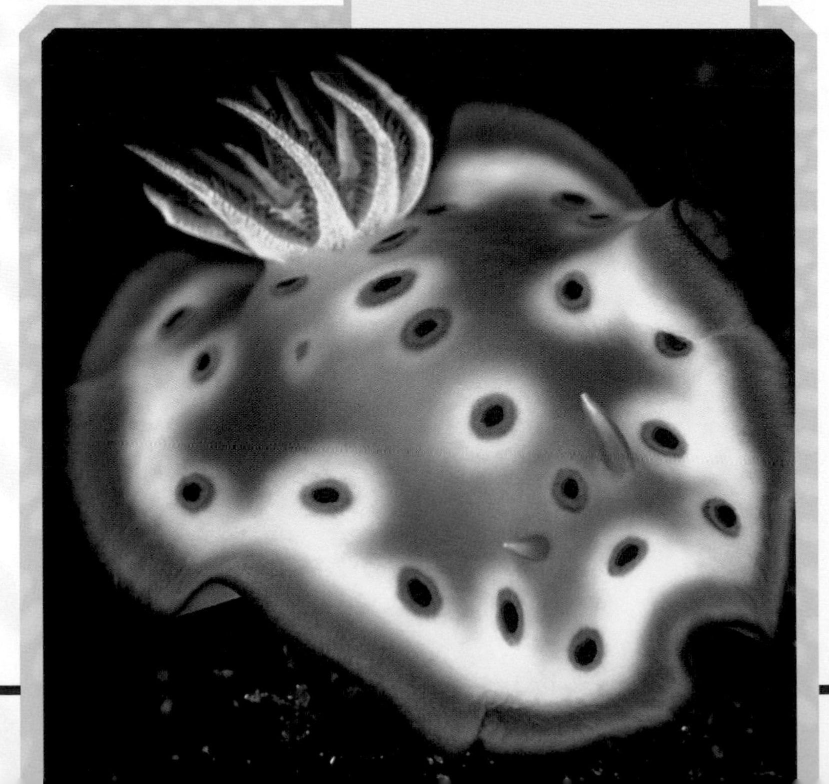

SCIENCE PIONEERS
Asa Gray and the plants of North America

Asa Gray was one of the founding fathers of **botany** in the United States. He had a network of collectors who went all over North America looking for specimens of different plants, which he then identified. He wrote a number of famous botanical textbooks, most famously the *Flora of North America*. He also wrote botany books for children to interest the next generation of plant collectors!

Asa Gray travelled widely and was well known in Europe. In fact, he was a close friend of Charles Darwin. In 1857, he became only the third person that Darwin told about his theory on natural selection, and Gray played a big part in introducing Darwin's ideas to the US. He died at the age of 77, respected around the world for his tremendous work in collecting and naming plants.

Asa Gray was always struggling to raise funding, but he managed to set up a herbarium at Harvard University in the US, which is still an important plant collection in the 21st century.

THE NAME GAME

People in different areas often have different names for the same thing. When everyone stays in their local area, this does not matter. But as people travelled around the world, and more and more living organisms were collected and identified, a big problem arose. People from different countries spoke different languages, so plants and animals were being discovered several times over and being given different names. When one scientist wanted to discuss an organism with a scientist from another country, how could they be sure they were talking about the same thing? Naming living things became a real issue, until the work of one great scientist solved the problem for good.

CLASSIFYING THE WORLD

In the 17th and 18th centuries, new types of animals and plants were being discovered all the time, and the way they were grouped and named depended on where they were found and who found them. Into the excitement and chaos stepped Carl Linnaeus, a Swedish botanist with a great love of plants, of teaching, and most importantly, of organizing!

Linnaeus qualified as a doctor but his real love was botany. He travelled on a number of expeditions and discovered many new specimens of plants. But his real genius was in organizing the specimens he found into groups. He based these groups on features of the plant, which could be observed by anyone. This was known as classification. In his earliest classification he used the arrangement of the sexual organs of the plants to put them into groups. He aimed to put every living thing into a few different groups, and create smaller groups within the main groups, ending up with a very small group, which he called the species. The beauty of his idea was that everyone could easily use the same system.

Carl Linnaeus is remembered for creating an efficient way of grouping and naming the living world, but he is not very easy to name himself. As well as Carl Linnaeus, he is also known as Carl von Linné and Carolus Linnaeus – take your pick!

In 1735, before he was 30 years old, Linnaeus published the first volume of his famous classification of living things called the *Systema Naturae*. This first volume was a slim pamphlet, but he continued to develop his ideas throughout his life and the *Systema Naturae* ended up as a huge book with many volumes.

WHAT'S IN A NAME?

Linnaeus' other great idea was to give every living thing a name that showed both the larger group of organisms it belonged to, and the exact species. Because he gave organisms two names it became known as the binomial system. Linnaeus used Latin as the language for his binomial system. This ancient language was widely understood by educated people all over the world, whatever language they normally spoke. Choosing Latin also avoided arguments about which language was chosen, and the problems which might have arisen if he had insisted on using his native Swedish.

The system that Linnaeus developed is still in use today. The features used to decide the groups have changed over the years, but the basic principles are the same. Every living organism is given a very long name in Latin. This name describes the different groups into which the organism fits, but it is the final two words in the list – the binomial name – that are used by scientists all over the world. So humans are known as *Homo sapiens*, honeybees are *Apis mellifera* and maize (sweet corn) is *Zea mays* (see page 37).

Classifying the living world

For centuries, scientists have been collecting animals and plants from all over the world and sending them to the great museums of natural history to be formally identified and classified. The Natural History Museum in London began life in 1753 as part of the British Museum, and it holds around 70 million specimens of animals, plants, and rocks. Although many of the species held there are now extinct, they have been identified, classified, and kept in a secure place, giving us a record of their existence. Similarly in the US, the National Museum of Natural History at the Smithsonian (founded in 1910) now holds over 125 million specimens which include plants, animals, fossils, minerals, rocks, and meteorites. These great collections, and others like them, contain the basis of our classification systems for all living things.

For around 250 years, people in the Natural History Museum in London have worked to identify and classify the living world around us. Joined more recently by similar museums in countries like the US, the work of these great institutions is very important in the 21st century.

HOW DOES CLASSIFICATION WORK?

Scientists classify the living world by dividing it into groups, and all the organisms in a group share certain characteristics in common. The largest groups are called **kingdoms** and they contain a huge variety of living things. The smallest groups are called species and they contain only one type of organism. Linnaeus divided the living world into two main groups – the plant kingdom and the animal kingdom – and this model lasted for more than two centuries.

However, by the 1960s, it was becoming clear that this model was too simple. Lots of organisms had been discovered which did not fit neatly into either of the two kingdoms. In 1959, a scientist called R.H. Whittaker suggested that it would be more useful to divide the living world into five kingdoms; animals, plants, fungi, **monerans**, and **protists**. This new model made sense, and was accepted quite quickly. It is still commonly used today.

SCIENCE PIONEERS Carl Woese

Carl Woese, a professor at the University of Illinois in the US, has suggested another way of classifying organisms. Instead of five kingdoms, he suggests the living world is divided into three **domains**, putting all the **eukaryotes** (organisms with cell nuclei such as plants, animals, protists, and fungi) into one domain called the **Eucarya**, and dividing the monerans (organisms without cell nuclei) into the **Bacteria** and the **Archaea**. This new system organizes living things by the way they have evolved, and makes **micro-organisms** very important. The work of Carl Woese is beginning to be accepted by biologists around the world.

A LOOK AT THE FIVE KINGDOMS

In spite of the work of Carl Woese, the five-kingdom model of classification is still widely used. An organism is placed into one of the kingdoms based on its characteristics.

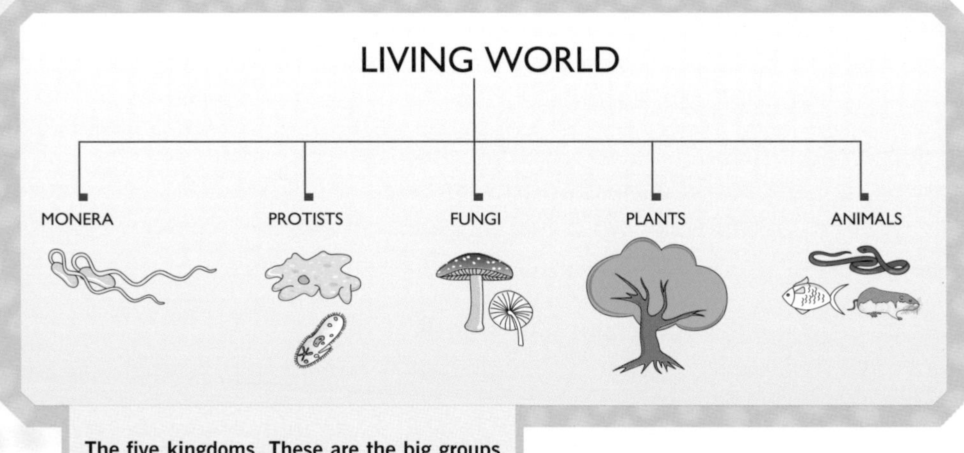

The five kingdoms. These are the big groups that are most commonly used to organize the living world – at the moment!

The monerans have the simplest cells of all living things. They are very small and they do not contain a nucleus. They are also known as **prokaryotes**. The organisms that make up the other four kingdoms all have a nucleus in their cells and are called eukaryotes. Bacteria and blue-green **algae** are monerans. There are about 4,760 known species of monerans, but it is likely that there are many more yet to be discovered.

The protists are mainly single-celled organisms and they include **protozoans** and some of the algae. So far 30,800 species have been discovered!

The fungi range from single celled yeasts to large toadstools. They have cell walls that do not contain **cellulose** and they cannot make their own food. Over 300,000 species have been found so far. But it is estimated that there are probably more like 1.5 million types of fungus in the world.

The plants all have more than one cell. They have cellulose cell walls and they contain green chlorophyll. Plants make their own food using photosynthesis and 422,000 species of plants have been discovered so far. However, scientists think that there are at least as many again still to discover.

The animals also have more than one cell, but unlike plants they do not have a cell wall. They cannot make their own food. Animals have a nervous system, and most of them move around to find their food. About 1,500,000 species of animals have been found so far – and there are many more to come!

ORGANISMS IN GROUPS

Once an organism has been put into a kingdom, it will be organized into a whole range of other groups. As the groups get smaller, the organisms in the group get more and more similar.

Kingdom – the biggest group which contains many similar phyla.
Phylum (plural phyla) – similar classes are grouped into a phylum.
Class – similar orders are put into a class.
Order – similar families are put into an order.
Family – similar genera are put into a family.
Genus (plural genera) – similar species are grouped into a genus.
Species – the smallest group, which identifies an organism completely.

Group	Human being	Maize plant	Honeybee
species	*Homo sapiens*	*Zea mays*	*Apis mellifera*
genus	*Homo*	*Zea*	*Apis*
family	*Hominidae*	*Poaceae*	*Apidae*
order	*Primates*	*Cyperales*	*Hymenoptera*
class	*Mammalia*	*Monocotyledoneae*	*Insecta*
phylum	*Chordata*	*Angiospermophyta*	*Arthropoda*
kingdom	*Animalia*	*Plantae*	*Animalia*

AN INVISIBLE WORLD – THE PROTISTS

It is easy to talk about the variety of life and ways of classifying living organisms in a very general way. But to really start to understand the enormous range of living organisms, it is important to look at some of the groups in more detail.

Everyday life is full of living organisms. The ones you notice are the big ones – trees, birds, dogs, cats, and so on. But everywhere you go there are millions of organisms you never see. They are so small that it takes a microscope to make them visible to the human eye. The protists are a good example of an almost unknown but very beautiful kingdom. They include plant-like cells that can swim, cells shaped like bells or fans or shells, and cells that live in great colonies. The protists make up a large part of the plankton in the sea, so they are a source of food for thousands of animals.

Tiny but lovely, this *Dendrocometes* protozoan is just one example of the protists which are found in fresh water and the seas all around the world.

SHINING SEAS AND RED TIDES

In some parts of the world, the sea at night gives off green sparks where the waves break or a swimmer splashes. This is caused by protists called dinoflagellates, which produce natural, biological light called bioluminescence. When there are a lot of these protists in one place, the sea literally glows.

Other dinoflagellates are not so attractive. Some types appear in large quantities and turn the seawater red. The red water contains poisons made by the protists, which can affect shellfish, fish, and even people. Red tides are not a welcome sight.

FORESTS OF THE SEA

Not all protists are small. Green, red, and brown algae are also members of the protist kingdom and they can be very big indeed. The best known form is seaweed, which exists in enormous colonies of cells. Kelp is probably the biggest protist of all. Kelp forests can be up to 60 metres (197 feet) tall!

DEADLY PROTISTS

There are a few protists that are well known because of the damage they cause. A number of terrible diseases affecting people, animals, and plants are the direct result of infection by protists. Dysentery, sleeping sickness, and malaria kill millions of people every year – each one caused by a protist. The Irish potato **famine** of the mid-19th century was also the result of infection by a protist that attacks potatoes and causes potato blight. The tiny organisms caused the deaths of a million Irish people, and led to more than a million emigrating to the US. Every kingdom in the living world has "goodies" and "baddies", and the protists are no exception.

THE WORLD OF PLANTS

There are about 422,000 known species of plants in the world, and quite probably as many again that have not yet been discovered. Plants are vital for life on Earth as we know it. In the process of photosynthesis, they use up carbon dioxide and produce the oxygen needed by all organisms for **respiration**. What is more, without plants to eat, animals would not be able to survive.

Most people have a mental picture of plants – green leaves, flowers, in a home, a garden, a park, a forest or the countryside... But the plant kingdom is very varied and full of surprises!

PLANTS WITHOUT FLOWERS

The plant kingdom can be split into two main groups – plants with flowers and plants without flowers. Plants without flowers have been around for a very long time. They were around long before the dinosaurs walked on the earth. Plants that do not flower reproduce by forming minute **spores**. They include many different phyla.

The mosses and liverworts (known as the bryophytes) are the simplest land plants. They do not have veins to carry water, which means they have to live in damp places so they do not dry out. The very largest member of the group is only 60 centimetres (2 feet) tall! Most bryophytes are found in tropical rainforests, where the warm, damp atmosphere suits them perfectly. However they live in lots of other places as well.

The ferns (known as the filicinophyta) are another well-known group of non-flowering plants. They have true roots, stems, and leaves so they do not need to live in damp places as much as the bryophytes. Also they can grow to be quite big. Ferns reproduce by forming spores on the back of their large, feathery, leafy fronds.

SEED-BEARING PLANTS

The great majority of flowering plant species reproduce using seeds. The conifers (trees like the pine and the fir) produce their seeds on special leaves, which develop into cones. All the other seed-bearing plants have flowers and fruit.

There are two main types of true flowering plants. **Monocotyledons** (usually referred to as monocots) have a single tiny leaf inside their seeds. They are very important for human life, because the grasses that feed our cattle and sheep, as well as our cereal plants, such as corn, wheat, rice, and barley, all fall into this group. They tend to have long, thin leaves with the veins running in parallel lines.

The **dicotyledons** (usually known as dicots) have two embryo leaves inside their seeds and their leaves have branched veins. They are the plants with the most variety around the world. Both the largest and the smallest known plants fall into this group.

There are plants from many different groups in this picture – yet it shows only the tiniest fraction of the variety of plant life here on Earth.

THE ANIMAL WORLD

The majority of the species alive on Earth today are animals. The number of species is estimated at anything from the 1.5 million already discovered to around 12 million! Animals cannot make their own food. They all rely on eating other living organisms, whether plants or other animals. Animals come in the most amazing variety of shapes, sizes, and lifestyles. Here are just a few of them...

ANIMALS WITHOUT BACKBONES

More than a million of the known species of animals do not have a skeleton on the inside of their bodies or any form of backbone. These animals are known as the invertebrates. Invertebrates include spiders, worms, starfish, jellyfish, **molluscs**, and insects.

The Cnidarian phylum contains some of the strangest and least animal-like animals – jellyfish, sea anemones, and **corals**. They all have circular, jelly-like bodies and most of them are **carnivores**, relying on special stinging cells to paralyse and kill their prey. A sting from the box jellyfish can kill a person in 4 minutes! Tiny coral polyps live in massive **reefs** and they build up over centuries. In contrast, some jellyfish travel thousands of kilometres in the ocean currents.

An octopus, a mussel, and the slugs and snails in your garden might not seem to have much in common, but they are all molluscs, the second biggest group of invertebrates. They all have a muscular foot on one side of their body and a hump containing all the organs of the body. This is often covered by a shell.

Worms come in a variety of types – round worms (known as nematodes), flat worms (platyhelminthes), and segmented worms (annelida) are all separate phyla – and there are huge numbers of them. They range in size from microscopic, such as the pinewood nematode, to the Giant Gippsland Earthworm which is one of the largest worms in the world. It is found in Australia, and it is almost one metre (three feet) long!

A SUCCESS STORY

Arthropods are the most varied and successful animals on Earth. About 900,000 arthropod species have been discovered and named. They have tough skeletons on the outside of their bodies and jointed legs and antennae. They have taken advantage of almost every possible habitat. The arthropod phylum includes crabs (crustaceans), spiders (arachnids), and the most successful group of all, insects.

THE INSECT CLASS

Insects make up the vast majority of species in the arthropod phylum. There are many more species which have not yet been described, so some scientists think there may be more like 10 million! Insects have very efficient muscles, and because they often change form during their life cycles – caterpillars to butterflies, larvae to wasps, and so on – they can eat several different types of food, which increases their chances of survival.

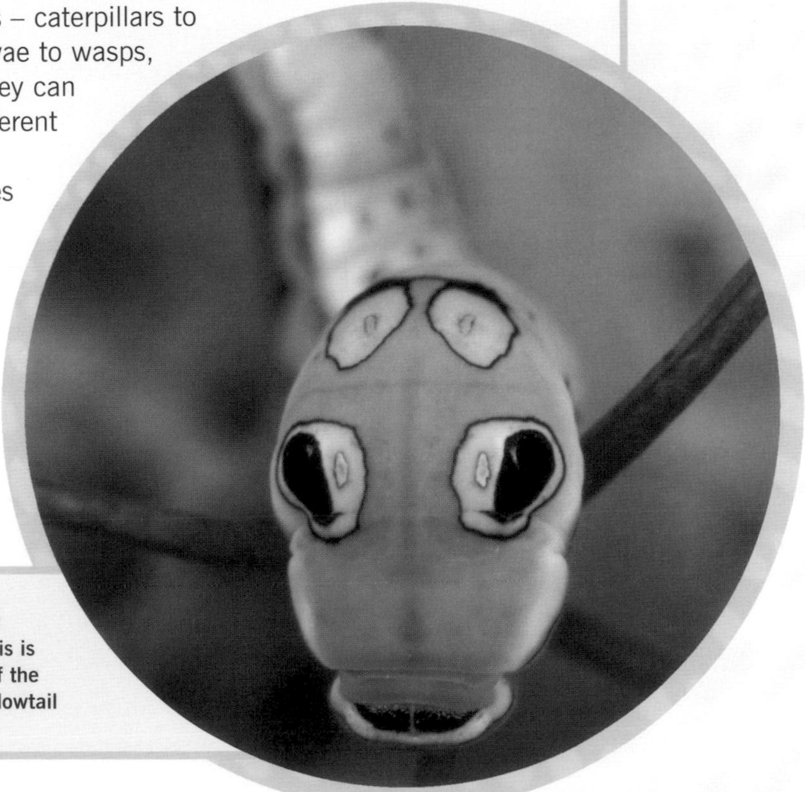

Insects have developed many exotic forms. This is the caterpillar of the Spicebush Swallowtail butterfly.

ANIMALS WITH BACKBONES

Animals that have a skeleton with a backbone inside the body are known as vertebrates, and they come in a wide variety of shapes and sizes.

WATER BREEDERS

Fish are a class of vertebrate with streamlined bodies for swimming and **gills** to take oxygen from the water, where they live their whole lives. Some fish (the sharks and rays) have rubbery skeletons made of cartilage, but 95 percent of fish species have skeletons made of bone and are therefore known as bony fish (for example salmon, cod, and minnows).

Fish are not the only vertebrates that need water to survive. **Amphibians** include animals such as frogs, toads, and newts. They have simple lungs and can move around on land, but they must return to the water to breed. Their eggs are fertilized outside their bodies, and the young animals (tadpoles) have gills and cannot breathe in the air.

MASTERS OF THE LAND

Reptiles (snakes, lizards, tortoises, turtles, crocodiles, and alligators) lay eggs with leathery shells, which allows them to reproduce on land. They have dry skin with scales, and lungs for breathing air. They can live for a very long time. The longest-living land animal was a tortoise given to the Tongan royal family by Captain Cook in the mid-1770s. It died in 1965 aged at least 188 years old!

RECENT DEVELOPMENTS
The world's smallest lizard?

In 2001, Blair Hedges and Richard Thomas, American biologists, discovered a new lizard species on a tiny Caribbean Island. It was the fiftieth new species they have discovered – and also one of the smallest known, as the tiny lizard (*Sphaerodactylus ariasae*) is only about 16 millimetres (0.6 inches) long!

MASTERS OF THE AIR

Birds have very light skeletons, beaks, wings, and feathers, and many of them can fly. They lay eggs with hard shells, which they then keep warm until the babies hatch out. The ability to control their own body temperature (this is called being **endothermic**) means that birds can stay warm and active even when the weather is cold.

THE MAMMALS

Mammals, like birds, are endothermic, and this class includes humans. They produce live young and, even more importantly, produce milk from special glands to feed their young after they are born. As a result, mammals can live almost everywhere – on land, in the water, and even, in a few cases, in the air. Mammals can live in the hottest and the coldest areas of the Earth and range in size from the smallest shrew, which weighs only a few grams, to the massive blue whale, which weighs in at around 140 tonnes.

This giant anteater is a mammal with a very striking feature. It has a tongue about 60 centimetres (2 feet) long that it uses to grab insects from inside their nests.

The species concept

Classification – the idea of grouping living organisms together and identifying which species they belong to – is very important in biology. Scientists estimate how many different species there are on Earth. People worry about species becoming extinct. Counting the number of species living in an area is a very important measure of **biodiversity**. But what exactly is a species?

LOOK AND SEE...

When Carl Linnaeus first introduced the idea of species, classification was based on looking at an organism and noting down its characteristics. Features such as the arrangement of the parts of a flower, the pattern of hairs on the legs of an insect, or the colours seen on the feathers of a bird were all used to group living things together into different species and then to put the species together to form bigger groups.

Making detailed observations of animals and plants has been part of identifying a species for many years. But in the 21st century, looks are not everything.

To begin with, the system worked well. But as more and more living organisms were discovered and new ideas like those of Charles Darwin came along, the old, rather simple picture of a species began to change. It was no longer enough to look at an animal or plant and decide which species it belonged to based on its appearance.

CAN YOU BREED?

The old way of defining a species has been replaced by something known as the biological species concept. This says that a species is a group of closely-related organisms that are able to breed to produce healthy, fertile offspring. It is a widely used concept. For example, human beings from different parts of the world can look very different and lead very different lives – think of an Inuit and a Masai – but this is not a barrier to having babies together because all humans belong to the same species. People often confuse grass snakes and vipers as they look similar, but they cannot interbreed at all. On the other hand, horses and donkeys look quite similar, and can actually interbreed to produce a mule, but mules are **infertile** because horses and donkeys belong to different species.

There are exceptions to this rule, particularly when it comes to plants, but on the whole it works for most of the larger organisms.

SCIENCE PIONEERS Ernst Mayr

Ernst Mayr was one of the greatest evolutionary scientists of the 20th century. The biological species concept was his idea. He was born in Germany in 1904, but emigrated to the US. He became a curator at the American Museum of Natural History in New York, where he worked on the classification of species of birds. At the same time, Mayr came up with his key ideas about **evolution** and species, which he published in 1942 in his book *Systematics and the Origin of Species*.

TAXONOMY TODAY

Classifying organisms into groups has become a science in its own right, called taxonomy. The process of deciding where an organism belongs in the world has changed a lot since Linnaeus first attempted to put animals and plants into groups. Modern **taxonomists** have a whole range of modern technology to help them get it right!

Obviously, if scientists can observe a group of organisms, they can make sure they interbreed and have fertile offspring. But this is not always possible, particularly if the animals or plants live in a very specialized habitat or have a very unusual diet so they cannot be kept in captivity. Another thing scientists can do is look at the way the embryos develop. Some organisms look very different as adults, but are very similar in the way their embryos grow.

RECENT DEVELOPMENTS
Witness for the Whales

A team at the University of Auckland in New Zealand has set up a project called Witness for the Whales, which aims to analyse DNA sequences from all the known types of whales. Anyone who works with whales anywhere in the world, can run a DNA analysis and then compare it with the results of the Auckland team to find out exactly which type of whale they have found.In some cases, they can even discover which particular population the whale has come from.

These magnificent mammals are some of the largest and most spectacular animals on Earth. DNA testing will help to identify them more easily.

Comparing organisms by looking at the way their body chemistry works can be very helpful. Some living things have very unusual chemical pathways. Some may share the same enzymes. Taxonomists can also look at the DNA of an organism. As it becomes easier to analyse DNA, this will become the main way of identifying species in the future. It is particularly useful for identifying single-celled organisms, which can be difficult to classify in other ways.

VARIETY IS THE KEY

Over the years, taxonomists have built up detailed descriptions of all the known species. These descriptions can then be used to help other scientists identify a particular species using an **identification key**. These can be very complicated, but you can get an idea of how to use one by looking at the simple example shown below.

If you do not know what a particular animal is, an identification key like this will help you to find out. Answer a series of questions based on what you can see until you have identified each type of animal.

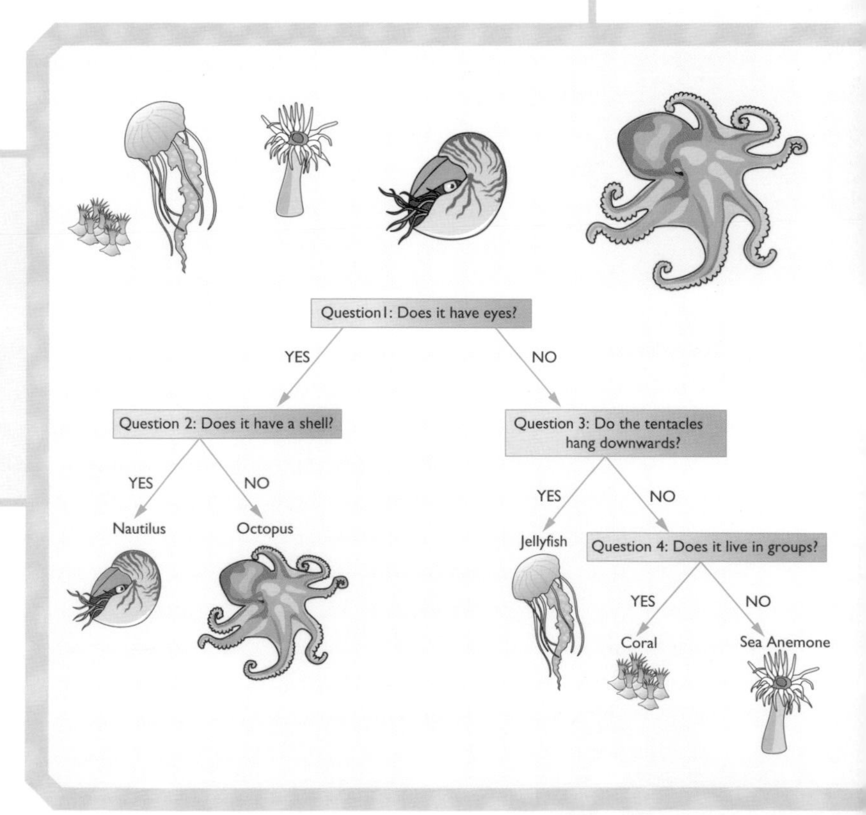

SPECIES THROUGH TIME

The variety of life on Earth is immense. There are millions of species of living things. But this great variety is nothing compared to the more than 4 billion species that scientists believe have lived on Earth throughout its history. The question of how all these different species have appeared and, in many cases, disappeared again was answered in the 19th century by two men, Charles Darwin (see page 19) and Alfred Russel Wallace. It is Charles Darwin who is remembered for the idea of evolution. However, Alfred Wallace had similar ideas and he too deserves a place in the history books.

The central idea behind the theory of evolution is that all living organisms have resulted from a long process of natural selection. Reproduction always produces more offspring than the environment can support. Those that are best adapted to their environment – the fittest – are the ones that will survive and breed, passing on their successful characteristics. Darwin's theory was that all of the species on earth are the result of this gradual process of evolution, of changing slowly over time, which he called survival of the fittest.

SCIENCE PIONEERS Alfred Russel Wallace

In 1858, Alfred Russel Wallace gave Charles Darwin the shock of his life. Wallace had gone on an expedition to South America with a wealthy amateur insect collector some ten years earlier. Unfortunately, when he returned home to England, the whole collection was destroyed in a fire. Wallace used the insurance money to go on another trip, this time to Borneo. When he was there, Wallace began to work on the idea that the animals and plants that were badly adapted to life were likely to die, leaving only the better adapted to survive and breed. He sent his ideas to the one man he thought would understand – Charles Darwin. Darwin was horrified! After working on his theory for so many years, he feared that someone else, who was less well known and had far less evidence, was going to publish the new ideas before he did.

In fact, Darwin and Wallace both published papers at the same time and nobody was impressed. But Darwin set to work immediately on his great book. He knew he had far more evidence than Wallace to back up his ideas, and he was determined to be first into print. Wallace and Darwin remained in contact for many years. They got to know each other well and enjoyed swapping ideas. Without the push from Wallace, Darwin might never have got round to publishing his great work.

Alfred Russel Wallace came from a poor family – unlike Darwin – but in spite of a difficult start he was a very gifted naturalist. By 1858, he was thinking along very similar lines to Darwin, and it was his work that pushed Darwin into action.

HOW ARE SPECIES FORMED?

The study of the history of life on Earth and the study of species alive today overlap in many ways. Findings about one help scientists to understand the other. Some of the clearest evidence for evolution is in the fossil record. Fossils are the remains of plants and animals from many thousands or millions of years ago, which are found in rocks, ice, or tar. Although the fossil record is not complete, it can give us a fascinating glimpse of life millions of years ago. It also shows us how some species have survived almost unchanged through time. Sharks, horse-shoe crabs, and horse-tail ferns, look the same today as they do in rocks millions of years old. On the other hand, we can also see how modern horses have evolved from tiny ancestors who emerged not long after the reign of the dinosaurs.

Fossils like these dinosaur eggs offer us a fascinating glimpse of the past, but we can never know the whole story. What terrible event prevented them from hatching?

FORMING A NEW SPECIES

Ninety-nine percent of all the species that have appeared on Earth have died out, but how exactly does a new species emerge? Remember the definition of a species – it is a population of similar organisms which can interbreed with each other. The most common way for a new species to be formed is when groups of individuals within a species become separated from each other for a long time. Each group changes over time due to natural selection, until they meet up again and they can no longer interbreed successfully. They have become separate species. There are several different ways this separation can happen:

- Physical isolation. This can be on a big scale, for example when the continents moved apart, or on a small scale, as when a new road cuts through an area where a species lives.
- Behavioural isolation. Changes in the breeding behaviour – when groups of animals are separated, their courtship display or the breeding season may change in response to things such as different light levels or different diets in the places where they are living.
- Mechanical isolation. Sometimes mutations can change the shape or arrangement of an organism's sex organs, making it impossible to interbreed with other members of the species any longer.

WHERE DID ALL THE SPECIES GO?

About four billion species have existed through the history of the Earth, but only a fraction of that number are alive on Earth today. At each stage, the animals and plants that were alive were the best adapted for the conditions at the time. As conditions changed, groups adapted to cope with the conditions and new species continued to emerge.

Biodiversity

There is a huge amount of variety both between and within species. This variety is important for species to survive in our changing world. A way to measure the variety of organisms in a particular area is to look at the biodiversity. The term biodiversity can mean anything from the total genetic variety in every organism everywhere to the number of different species found in a particular habitat. In the 21st century, biodiversity is taken very seriously. Lots of different species, and lots of variety within each of those species, is seen as a very good thing.

WHY DOES BIODIVERSITY MATTER?

In many areas of the world, habitats are changing rapidly. Sometimes this is the direct result of things that people have done – building a dam or felling trees in a rainforest. Sometimes it is a result of natural events – floods, forest fires, volcanoes, or earthquakes. Whatever the cause of the change, if there are lots of different species (or high biodiversity), there is a greater chance of an area recovering and plant and animal life continuing. This is because it is more likely that at least some of the species will have adaptations that make it possible for them to survive the change. In areas with little biodiversity, change can mean many species become extinct.

Biodiversity matters a great deal from a human point of view, too. We rely on other organisms for many things, especially food. Around the world we use around between 30,000 and 70,000 different species to feed ourselves. Insects are vital to pollinate many of our crops. Thirty percent of our medicines were originally discovered in plants. If biodiversity is lost, we risk losing these benefits – the ability to improve our food crops and the chance to discover new medicines. We will also be damaging the living world around us which supports us so well.

RECENT DEVELOPMENTS
measuring biodiversity using lizards

In sub-Saharan Africa, sleeping sickness is carried by tsetse flies and kills hundreds of thousands of people and cattle. In Zimbabwe, trees are sprayed to kill the flies, but this kills other creatures as well. Lizards eat these insects. If the number of insects falls, so does the number of lizards. In turn, lizards are eaten by birds. If lots of lizards are seen on the trees, it means that all sorts of insect and bird populations are doing well and biodiversity is good. But if numbers drop after spraying, it means biodiversity is falling as more than just tsetse flies are killed.

These lizards in Zimbabwe can be used as bioindicators – animals or plants that let us measure what is happening in a habitat.

EXTINCTION MEANS FOREVER

Only 1 percent of the species that have ever lived on Earth are alive today; the other 99 percent are extinct. Throughout the history of life on Earth, new species have been evolving. At the same time, older species that cannot cope with the new competition have died out. There have been five different occasions during the history of the Earth when extinctions have taken place on an enormous scale (**mass extinctions**).

Extinction happens as a result of change and variation, and it is part of the process of evolution. However, many scientists are beginning to get very worried. Mass extinctions of the past took place over millions of years, but now species are becoming extinct faster than ever before. Some scientists predict that by the year 2010, a quarter of all the species currently on Earth will have been lost.

The Pyrenean ibex lived in the mountains of Europe. They became extinct in 2000, despite efforts to save the species. Scientists took samples of ear tissue from the last ever ibex in the hopes of cloning some replacements in the future.

No-one knows just what the effect of these rapid modern losses will be. There may be a knock-on effect. As one species becomes extinct, it can affect the survival chances of another, and the loss of so many different organisms could make the whole **ecosystem** of the planet unstable.

HOLDING ON TO WHAT WE'VE GOT

Around the world, people are becoming more aware of the importance of biodiversity, and the problems of the natural world. Many scientists believe that the climate of the world is changing as a result of the pollution produced by people. Burning **fossil fuels** produces many poisonous chemicals that can damage the environment. In lots of places, people are working hard to reduce the damage, although unfortunately it is often the biggest producers who are slowest to take action. However, all around the world, people are trying to help. Better, sustainable farming techniques are being developed, hedges are being replanted, areas where species are under threat are being protected, and schemes to increase biodiversity are being set up.

There are probably still millions of species on Earth that have not yet been classified. It is likely that many of them will become extinct before they have even been discovered. Many mammals and birds are under threat in Indonesia, India, Brazil, and China, while plant species are disappearing fast in Southern and Central America, Central and West Africa, and Southeast Asia. People are trying to stop the losses, but only time will tell how many species will survive and how many will be lost forever.

Hope for the future

Here are just a few of the ways in which people are trying to help preserve the biodiversity of our planet.

CHROMOSOME ANALYSIS

DNA analysis of living organisms makes it possible to identify new species quickly. Japanese scientists analysed DNA samples from a number of dead whales which they thought were rather small, thin, fin whales. The DNA showed that they were in fact a previously unknown species!

DNA can also be used to help save species. In the US, the wild condors of San Diego were dying out. All the known birds were captured in 1985, and DNA analysis showed that one of the three remaining groups of birds had a genetic disorder that meant most of the babies died before they hatched. A specially designed breeding programme was set up and in twenty years, the condor population was up to 215. In the last few years, around 90 condors, without the faulty genes, have been released back into the wild.

SOWING THE SEEDS OF SUCCESS

Plant species are also under threat all around the world, but scientists are fighting back. In a massive international project masterminded from the Royal Botanic Gardens at Kew in London, a plan to protect 24,000 plant species by storing their seeds has been developed. In the UK, seeds from all the wild flowering plants have been collected for the Millennium Seed Bank Project (MSBP) and put into storage. These UK seeds are not alone in their storage jars – seeds from around the world are sent to join them!

Sixteen countries from Jordan to Madagascar, and from Botswana to Mexico, are involved in the Millennium Seed Bank Project. In the US, the project is called "Seeds of success". Scientists aim to collect seeds from all of the remaining 1000 species of prairie tallgrasses and save them for the future.

South Western Australia is one of the world's top "hot spots" for biodiversity. It has many plants – at least 12,000 species are known! Seeds from all of the threatened species are being collected and stored, both in Australia and in the UK.

Projects like these are helping to classify the huge variety of life on Earth and to protect that variety for the future. Scientists, governments, and individuals are all working towards protecting and even increasing the biodiversity of the planet. For all our sakes, we must hope they are successful!

Once the seeds for the MSBP have been collected they are checked, dried and then stored at about -20 °C (-4 °F). Under these conditions, they can survive for decades, ready to be used to reintroduce species as habitats are protected.

Further resources

MORE BOOKS TO READ

Goodman, Polly, *Animal Classification: A Guide to Vertebrates* (Hodder & Stoughton Childrens, 2004)

Stockley, Corinne, *The Usborne Illustrated Dictionary of Biology* (Usborne Publishing, 2005)

Townsend, John, *Incredible Creatures series* (Raintree, 2004)

Nature Encyclopedia (Dorling Kindersley, 1998)

USING THE INTERNET

Explore the Internet to find out more about variation and classification. You can use a search engine, such as www.yahooligans.com or www.google.com, and type in keywords such as *species*, *chromosome*, *mutation*, *taxonomy*, or *biodiversity*.

These search tips will help you find useful websites more quickly:

- Know exactly what you want to find out about first.
- Use only a few important keywords in a search, putting the most relevant words first.
- Be precise. Only use names of people, places, or things.

Glossary

alga simple plant containing no true stem, root, or leaf

allele form of a gene

amphibian type of animal that lives both in and out of water, such as frogs, toads, and newts

Archaea suggested domain for the classification of unusual prokaryotes

asthma condition that affects the respiratory system, that causes breathing trouble. In a worst case this can lead to death.

autosome chromosome that carries information about the body cells

bacteria type of micro-organism that can be helpful, but that can also cause disease

Bacteria suggested domain for classification of all traditional bacteria

base important building block of the DNA molecule

biodiversity measure the diversity of organisms living in a given area – both the different types of organisms and the variety within species

botany scientific study of plants

carnivore animal that eats only other animals

cellulose material that makes up the cell walls in a plant cell

chromosome thread-like structures found within the nucleus of cells, containing DNA

coral type of marine animal that makes a hard stony substance that forms an external skeleton

cystic fibrosis genetic disease that causes problems with the respiratory system

dicotyledon plant that has two embryo leaves inside its seeds

DNA (deoxyribonucleic acid) molecule that carries the genetic code. It is found in the nucleus of the cell.

domain major classification group suggested by Carl Woese

dominant dominant characteristic occurs even when only one allele is present in the allele pair

double helix shape of the DNA molecule, which is made up of two spirals fitted together

ecosystem all the animals and plants living in an area, along with the interactions between the living organisms and the things that affect them such as the soil and the weather

embryo baby at a very early stage of development inside the mother

endothermic animals that can control their own body temperature

enzyme protein molecule that changes the rate of chemical reactions in living things without being affected itself in the process

Eucarya suggested domain for classification of all eukaryotic organisms

eukaryote organisms with cell nuclei

evolution idea that new species develop by a process known as natural selection – the survival of the fittest

famine extreme lack of food

fossil fuels fuels formed over millions of years from the remains of ancient plants and animals. They are oil, coal, and natural gas.

fungi kingdom of organisms that do not move around and cannot photosynthesise

gamete sex cell (sperm or egg)

gene unit of information in the DNA

genetic disease disease that is inherited in genes from the parents

genome all of the genetic material that makes up the chromosomes of an individual

gill organ for extracting oxygen from water

habitat place where an animal or plant lives

haemophilia genetic disease that affects the blood

herbarium systematic collection of plants

identification key key to help identify a particular living organism using a series of questions

infertile unable to breed

inheritance process of receiving something from parents

kingdom name given to the five main groups of living organisms – the biggest groups in classification

mammals group of vertebrate animals that produce live young, and produce milk from special glands to feed the young

mass extinction occasion when large numbers of species of living organisms die off completely within a relatively short time

micro-organism bacteria, viruses, and other minute organisms which can be seen only using a microscope

molecule group of atoms bonded together

mollusc group of invertebrates that have soft bodies and hard external shells

moneran (see **prokaryote**)

monocotyledon plant that has a single tiny leaf inside their seeds

mutation change in the genetic material of an organism

natural selection survival of the fittest organisms, and the passing on of their genes through reproduction

naturalist someone who studies natural history

nucleus central part of the cell, the nucleus contains the DNA

ovum female sex cell in animals, also called an egg

photosynthesis process by which green plants make food from carbon dioxide and water, using energy from the Sun

placenta part in the uterus which connects to the developing foetus to provide food

pollinate transfer pollen from the male to the female parts of a flower

predator animal that hunts other animals to obtain food

prokaryote organism without cell nuclei

protein important building block of living things

protist microscopic living organism, mainly single-celled

protozoa single-celled microscopic animal

radioactive gives off radiation

recessive recessive characteristic only occurs if the allele pair is made up of two recessive alleles

reef ridge in the sea made of rock, coral, or sand

reptile group of vertebrates that lay eggs so they can reproduce on land. They have dry skin with scales and lungs for breathing air.

respiration using oxygen to release energy from food

sex chromosome chromosome that determines the sex of the new individual

sexual reproduction involves the joining of male and female sex cells to form an individual that is different from both its parents

species specific group of very closely related organisms whose members can breed successfully to produce fertile offspring

sperm male sex cell in animals

spore tiny reproductive unit, typically just one cell

taxonomist scientist who specializes in classifying different organisms

ultraviolet radiation beyond the light spectrum

unethical something that is not morally right

uterus part of the female reproductive system where a baby develops

X chromosome sex chromosome in humans that carries information about being female

Y chromosome sex chromosome in humans that carries information about being male

Index

Titles in the *Life Science in Depth* series include:

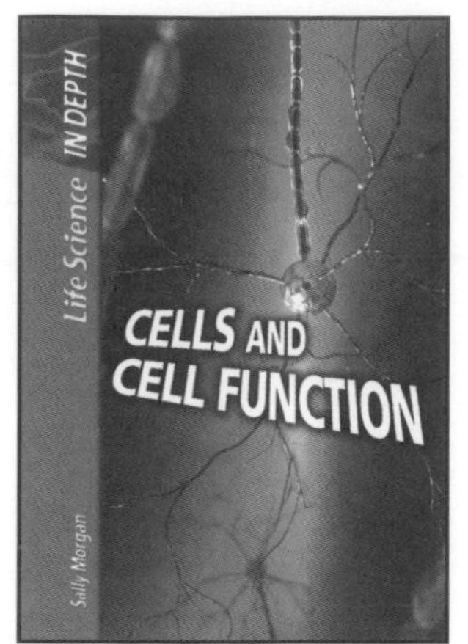

Hardback 0 431 10896 X

Hardback 0 431 10897 8

Hardback 0 431 10898 6

Hardback 0 431 10899 4

Hardback 0 431 10900 1

Hardback 0 431 10901 X

Hardback 0 431 10910 9

Find out about other titles from Heinemann Library on our website www.heinemann.co.uk/library